The C.Y. O'Connor ERADE Village Foundation

Incorporating the Immunogenetics Research Foundation and the Institute of Molecular Genetics and Immunology Incorporated

Education, Research And Development, Employment

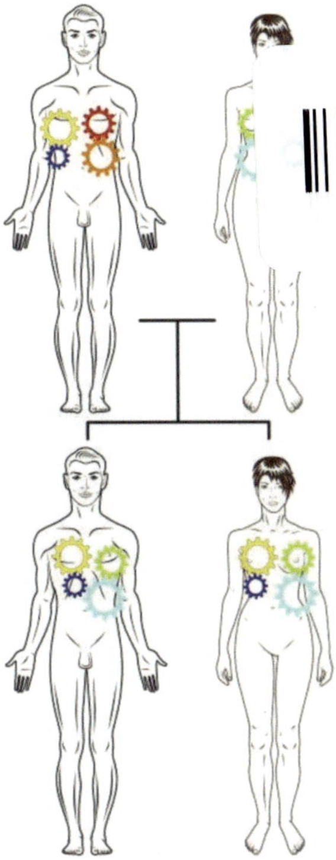

34th Annual Report 2017

C Y O'Connor ERADE Village Foundation Inc.
Employment Research and Development
Employment Technology Incubator

Street Address	1574 Hopeland Road Nambeelup WA 6207
Postal Address	PO Box 1 North Dandalup, WA 6207
Telephone	(+61 8) 9397 1556

Key Contact	Professor **Roger** Dawkins
Position	Chairman
Email	rldawkins@cyo.edu.au
Second Contact	Dr. Sally Lloyd
Position	Director
Email	slloyd@cyo.edu.au

Company Overview: CYO 'Connor ERADE Village Foundation seeks to harness the intellect of young Australian post graduates whilst maintain extensive collaboration with leading institutes overseas.

Recently expanding into multidisciplinary science in fields such as bioenergy, information technology, natural products, recycling and water, the core technology is interrogation of genetic information.

The aims include understanding of human evolution in the context of disease including retroviruses, identification of new genetic markers via genomic prospecting and development of diagnostic tests for inherited traits in numerous species of medical and commercial interest such as cattle, dogs, fish and invertebrates.

Product/Service Information: The CY O'Connor Education Research and Development Employment (ERADE Village) Foundation is a charitable Foundation that conducts independent and sponsored research and administers a federally funded technology incubator.

The Foundation has available low-cost two-story R&D studios including accommodation if required, central laboratory facilities, conference rooms, visitor accommodation and sophisticated intranet and broadband connection.

Fellowships are awarded to outstanding researchers to assist them in commercialisation of their R&D.

Industry and government funding have established the Village as a multi-million dollar R&D and commercial centre following a model that is being replicated at the Albany Research Institute

Published by Nearurban LLC, 1603 LBJ Freeway,
Suite 280, Dallas, TX 75234
Tel (972) 974 7130, Fax (214) 393-4001
Email: admin@nearurban.net

Manufactured and printed by CreateSpace.

ISBN: 978-0-9864115-4-0

Title and Cover image:

NU Publishing

FRIENDS OF CYO'CONNOR VILLAGE

The Foundation is a charitable organisation supporting education, research and development and employment. Without the help of friends of the Foundation it would not be possible to achieve all these objectives, including exciting new scientific discoveries.

The Board thanks its many friends for their generous donations, bequests, collaborations, advice and assistance during 2016.including: C.Y. O'Connor Village Pub for the donation of function facilities, food, beverage and service for Foundation meetings, Tillbrook Nominees Pty Ltd for the donation of 1/3 of all TAB commission, AL&MD Philanthropic Foundation for the donation of Fellowship and Project funds, Goldfather Corporation Pty Ltd for the donation of Administrative and Accounting Services, Melaleuka Stud for the donating the use of Nambeelup Research Facility, and Dom Bayard for donation of time and expertise.

I/we (Name/s) .... (Tel)...............

Of (Address)..

wish to:

Join the "Friends of CYO" 2017 ($55.00 pa) $................

Receive information about the Corporate Golf Day and other fundraising events ......

Be invited to Scientific Retreats ......

Receive information about arranging a bequest ......

Support the Foundation in other ways ......

Please accept my / our total payment of $................

Cheques/Money Orders payable to: The CYO'Connor ERADE Village Foundation.

OR

Please debit my **Credit Card** with the above total amount only. I understand that American Express, Diners, Mastercard or VISA are accepted. My card details are as follows:

Card Type: ....

Card No.: .... Expiry Date ...../.....

Cardholder's Name: .... Signature ....

POST TO: THE C. Y. O'CONNOR ERADE VILAGE FOUNDATION INC.
PO BOX 5100 CANNING VALE SOUTH WA 6155

AL & M DAWKINS FOUNDATION

Sponsors up to

$200,000 p.a. for Research & Development in WA

Residential scholarships and grants for:

FUTURE OF WA
BIOMEDICAL RESEARCH

in association with a university within WA if enrolled. Matching funding must be available or anticipated. Expressions of interest including track record should be addressed to:

CY O'Connor Village,
PO Box 5100, Canning Vale 6155

Closing Date

November 30th or May 31st each year

The AL & M Dawkins Foundation has been funded to support problem solving in Western Australia by:

(1) Defining important issues relating to urban development and planning and potential solutions thereto;
(2) Defining important issues relating to biomedical research and solutions thereto;
(3) Supporting prospective innovators in the above fields and by funding awards, scholarships, prizes and fellowships in association with CY O'Connor ERADE Village, or any branches thereof, in association with any university to which the CY O'Connor ERADE Village may be associated;
(4) Provide funding for patent costs to protect any intellectual property so developed.

Applications received by November 30 or May 31 each year will be evaluated by one of two expert panels which make recommendations to the Advisory Board,

The Board may seek further advice and will also consider:

(1) the relevance to multidisciplinary projects underway within the CYO
(2) the value of the individual to life within the village
(3) the capacity to leverage other funds
(4) university linkages
(5) potential IP

and will offer funding and accommodation for live-in researchers.

The town planning panel will be seeking to support innovative projects relating to water shortages and contamination, affordable housing, sustainable villages and isolated communities.

The biomedical panel will identify boutique projects which can be internationally competitive whilst relevant to local issues such as DNA matching, evolutionary genetics, livestock diseases and plant genomics.

Within 3 months, the administrators of the fund will advise applicants in the following categories

(1) successful
(2) for reconsideration when matching funding is confirmed
(3) invitation to reapply in the next cycle
(4) unsuccessful

Grants, generally of $50, 000 must include a stipend and rental but may also include funds for technical support, equipment, reagents, patents and publications. Travel grants will not be considered. The Foundation may decide to make no awards.

Funds will only be released after assignment of IP in return for a royalty negotiated on a case by case basis.

Table of Contents

Patron
Mr. S.L. Perron AM
Vice-Patrons
The Hon. Kim Beazley MA, MPhil
Senator Alan Eggleston BA, MRCS, LRCP, FRACGP
Mr. Max McDonald
Mr. Patrick Dawkins
Mr. Syd Martin

Board of Management

as of April 2016

Appointed Members

Professor Roger L. Dawkins (MD, DSc, FRACP, FRCPA, FRCP)	- Research Director/Chairman
Mr. Anthony Lloyd (BSc (Ag)(Hons)MBA,LLB)	Vice Chairman
Mr. Patrick Dawkins	- Deputy Chairman
Mr. Renny Lee-Steere (EMBA)	- Deputy Vice Chairman
Mr. Rob Giura	- Honorary Secretary
Mr. Ben Dawkins (LLB)	- Deputy Honorary Secretary
Mr. Wayne Cox (MTax, DipFS(FP), CPA)	- Honorary Treasurer
Mr. Jason Wallis (BBus, R.E.Lic)	-Deputy Honorary Treasurer

Elected Members
Mr. A Pitsikas
Mr. A Peggs
Mrs. L Ashdown
Mr. Brian Walker
Co-opted Members (by Invitation)
Dr. H. "Reb" Rebhan, (DVM)
Dr. Sally Lloyd (BSc, PhD)
Life Members
Dr. BJ Stewart, BVSc (Hons), Dip Ag
Mr. John Heydon, BCom, CA, FTIA
Mr. David GA Gibb, BA (Hons), AIBMS, AAIMS
Mr. Lindsay Wolridge
Dr. Edward J Steele

Executive Reports

Chairman's Report

Professor Roger L. Dawkins

CYOEVF Establishes Collaborative Research Hub with Murdoch University, Peel Development Commission and Local Farmers

The Foundation's purposes are evolving, in keeping with the relocation to North Dandalup and the associated commitment to improving agriculture and food production within the Peel region.

Who better to consult than the local farmers? It became obvious that some have succeeded through adaptation and now have the confidence to invest their own futures in problem-solving. We gather together for a sandwich each Wednesday and discuss the challenges. As a consequence, we now have a sufficient understanding of the problems to develop some practical solutions. Some are very clear even if daunting.

Lack of Skills

Increasingly, there is a shortage of skills, especially amongst those young and fit enough to contribute to long-term solutions. Successful farmers have always been multi skilled. Hence, the need for a multidisciplinary Education, Research and Development, Employment (ERADE) Village including an R and D Incubator and a training program as a means of recruitment. Previous experience teaches that an apparently irrelevant interest, or even hobby, can grow and develop into problem-solving. One strategy is to invite farming families to nominate and support their own trainees.

Water and Power

In spite of wet winters, there is a lack of water for agriculture. The rivers and creeks have almost disappeared. The locals suspect that this is largely due to bauxite mining over the Darling scarp. The level of the most superficial aquifer falls quickly over summer. Improved recharge of appropriate aquifers and recycling of waste water are partial solutions but do not appear to be happening. The high price of power (despite an excess capacity in the grid) is an impediment. Solar and wind are abundant and could provide the solution.

We will support the development of maintenance free, water efficient, modular solar irrigation, utilizing the superficial aquifer and improved storage of excess run-off. At present, all relevant innovation is imported which seems to increase the cost, especially of repairs and maintenance. Furthermore, solutions developed in Israel and Germany are not immediately applicable to the special features of the Peel region.

The existing pasture is not optimal. The challenge is to develop deep-rooted perennial mixtures which can chase the falling aquifer in summer, strip the nutrients for their own use and maintain nutritional value with only supplementary irrigation

Adapted livestock

Some of the locals hope to establish Peel as a source of livestock adapted to the many very different conditions throughout the state. Ruminants such as cattle and sheep, in contrast to humans, have the ability to convert cellulose to protein and have much to teach in attempts to produce cost effective and nutritious food. Some breeds like Wagyu not only convert to protein but also to the healthy fat which is so beneficial to humans. Strangely, the metabolic processes and the underlying genetics are still poorly understood but do appear to be amenable to collaborative research and development here within the Peel region.

A Short History of the CYOEVF

There are three phases:

1983 to 1998 - Fundamental discoveries in Immunology and Genetics

The Immunogenetics Research Foundation was established largely to respond to the challenges of AIDS. During the next 15 years, supported by National Health and Medical Research Council funding, we improved the use of diagnostics using Bayesian logic, we saw HIV in tissues, before others, and we discovered ancestral haplotypes which regulate immune responses such as those controlling HIV, transplants and autoimmune diseases. More recently, we have shown that similar haplotypes in Wagyu cattle regulate fatty acid metabolism and other commercially important traits.

1998 to 2013 -

The IRF became the CYOEVF with a focus on multidisciplinary incubation within the ERADE Village established on Melaleuka Stud in what is now the suburb of Piara Waters. With a mix of federal, state and charitable funding, 20 incubator units were built next to state-of-the-art laboratories within the old farm buildings and a demountable school house. During this period, most staff and incubees were supported by the proceeds of patents until they were able to establish their own funding.

2013 to 2023 -

With the necessary move of Melaleuka Stud to North Dandalup, the laboratory followed and the focus shifted to grass roots problem-solving in the Peel region. Now, we are committed to establishing a new ERADE Village and rejuvenating the established township with all its potential advantages including an existing community, the railway, and excellent north-south and east-west connectors. The intention is to take advantage of sufficient space to extend the scope and the life of Research and Development Incubation whilst allowing individuals to own a business within the commercial facility. The collaboration with Murdoch University permits access to excellent facilities at the various locations including Perth, Rockingham, Whitby Falls and Mandurah and introduces a whole-of-Peel approach. This is important since we aspire to strong linkages between interdependent sites including the noxious industries within Nambeelup. Also, because youth unemployment is a shared problem.

Currently, we are establishing a satellite laboratory for Global Reproductive Services in Victoria. The main focus is to screen Wagyu embryos before implantation, thereby improving the production of healthy (low melting temperature) intramuscular lipids. At the same time, using protocols developed for humans, it is possible to reduce recessive deficiency disorders. These approaches can also be established within the ruminant facility at Whitby Falls thereby permitting the export of superior genetics.
Murdoch has invested over $2m in refurbishing the facilities on the farm and is "open for business" when it comes to local collaboration in research, teaching and enterprise.

Big Data and Meta-Analysis

The current enthusiasm for "big data modelling" and for globalized meta-analysis contrasts with the ERADE Village approach. We see value in restoring a role for the development of ideas arising out of local observation and problem-solving. We are inspired by the views of Desmond J Sheridan and hear his warnings of the implications of the shift to retrospective analysis rather than hypothesis-driven discovery.

In *EVIDENCE-BASED MEDICINE: BEST PRACTICE OR RESTRICTIVE DOGMA* – available on Amazon- he says:

> *"This distinction between comparative research and science is ...important .... The past three decades have seen a great expansion of comparative research in healthcare and to a large extent this has been at the expense of science and clinical science in particular, ... this (change of emphasis) has been a major contributing factor to the decline in new clinical discoveries and the failure to translate new discoveries in biology into useful clinical benefits."*

As I emphasized last year the same decline is happening in agriculture in WA.
The challenge is to use "big data" such as those generated by genomics and metabolomics, whether locally or remotely, as part of the strategy to address hypotheses arising through problem-solving. Prototypes developed within the R and D incubator will be tested in field trials involving local farmers. The key is to engage young talents before they become entranced

with alternative cultures. Murdoch has the resources to store, analyse and interpret the vast quantities of data that can now be generated. It is one thing to collect the data, it is another to translate the information in to a usable form for farmers and scientists alike.

Collaborative Research Hub at North Dandalup

The Foundation is developing detailed plans for the hub with input from Peel Development Commission (Paul Fitzpatrick, Andrew Ward, David Arkwright), the Murray shire (Dean Unsworth, Rod Peake, Brett Flugge), Murdoch University (David Morrison, Chris Hutchinson, Richard Harper) and Tillbrook Melaleuka Group (my sons Ben and John). In the preferred solution, TMG provides land for each of the parties and for the shared facilities, PDC and Murray (with state and federal assistance) provide the buildings and support for local trainees and Murdoch provides the staff and postgraduate students who will be funded by CYOEVF to a cap of $1mpa. This collaborative approach will be used for research and the development of extension and pathway programs into the University. Murdoch has lifted the cap of student numbers going into Animal Science and has recently introduced a Crops and Pasture program into its educational offerings.

Acknowledgements

Thank you to the Board members for their patience in developing the plan and for their care in committing the Foundations resources. I have taken inspiration from Tony Lloyd and Herbert Rebhan in particular. On behalf of the Board and our partners, thanks also to Ben and John for their generosity in providing their land and their computer network to link the various Peel sites and farmers.

We all appreciate the funding provided by Wayne Cox and Renny Lee-Steere, as trustees for ALMDF, for support of crowd funding and for the additional site in North Dandalup. Sally Lloyd, Jose Valenzuela and Lindsay Baker have kept the existing research vibrant in spite of an excess of planning and the subsequent delay in execution. We are grateful to Pardoo Beef Corporation, Milne Feeds and the PDC for financial support. I thank Prof David Morrison, DVC, Murdoch University, for his contributions to this report.

R L Dawkins

Deputy Research Director's Report

Dr. Sally S Lloyd

Healthy Beef Research Program

This year saw the maturation and slaughter of the first F1 Red Wagyu steers from Melaleuka Stud. It quickly became apparent that their reputation of Akaushi Beef for healthier fat was not all hype. A comparison of the pure Simmental steers to the Red Wagyu crosses, produced under similar conditions, showed that the F1 Red Wagyu had significantly lower Tm. The very long feeding traditionally used with black wagyu was not required, differences could be seen with only 50-150 days of feeding. With additional genetic selection and optimization of feed our goal of beef with Tm below body temperature on less than 150 days will be achievable.

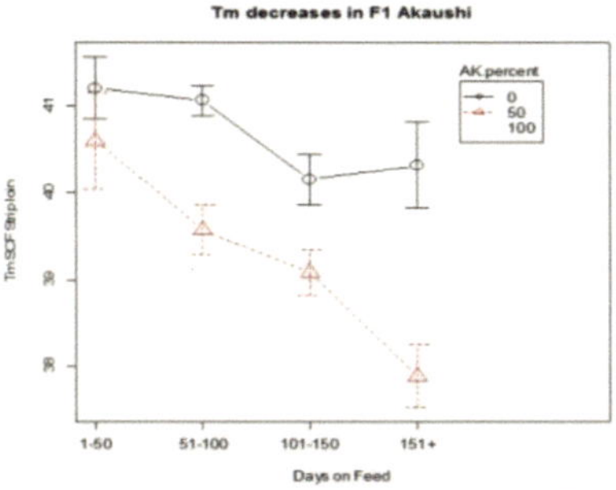

Breed	1-50 DOF	51-100 DOF	101-150 DOF	151+ DOF
Euro	25	85	38	10
F1 Akaushi	3	25	27	6

New "On Farm" Laboratory

At the beginning of 2016 the centre of CYO research activity moved to the newly finished laboratory on Hopeland Road, Nambeelup. This facility is now set up for DNA extraction, PCR, DNA marker analysis, Microscopy, Tm measurement and long-term sample storage. By the end of the year we realized the need for additional space, including a seminar room with capacity for 30 people.

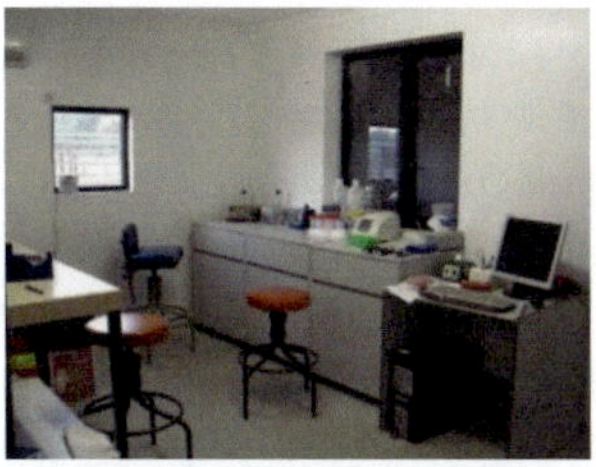

Patents

特　許　証
(CERTIFICATE OF PATENT)

特許第５９７６５４１号
(PATENT NUMBER)

発明の名称 (TITLE OF THE INVENTION)　家畜のジェノタイピング方法

特許権者 (PATENTEE)　オーストラリア国　６１５５　ウェスタン　オーストラリア，カンニング　ベイル　サウス，ウォートン　ロード　６２５
国籍　オーストラリア連邦
シーワイ　オコーナー　イレイド　ヴィレッジ　ファウンデーション

発明者 (INVENTOR)　ドーキンス，ロジャー　レッツ
ミルマン，ジョン　アンソニー
ウィリアムソン，ジョセフ　フレデリック
その他別紙記載

出願番号 (APPLICATION NUMBER)　特願２０１２－５２９０６６

出願日 (FILING DATE)　平成２２年　９月１７日

登録日 (REGISTRATION DATE)　平成２８年　７月２９日

この発明は，特許するものと確定し，特許原簿に登録されたことを証する。
(THIS IS TO CERTIFY THAT THE PATENT IS REGISTERED ON THE REGISTER OF THE JAPAN PATENT OFFICE.)

平成２８年　７月２９日

特許庁長官
(COMMISSIONER, JAPAN PATENT OFFICE)

小宮義則

We have put considerable investment into establishing the Foundation's patent portfolio in 2016, with some notable success.

Patents were granted for the "Methods of Genotyping Livestock" patent in Japan, Australia and United States. We are still pursuing the application for this patent in Europe.

National Phase Patent applications for "Methods And Systems For Measuring Melting Temperatures" have been submitted in Australia and the United States.

The Future

In 2017, I look forward to working with Richard Harper, Phil Nichols, Rob Trengove, and others from Murdoch University to expand the breadth of CYO'Connor Foundation research into. There are exciting opportunities to improve pasture and cattle feeding, apply the metabolomics to understand the fatty acid pathways in fattening cattle.

In particular, we plan to identify the genetic variations of endemic kikuyu varieties and select and improve on them. The end point will be a more palatable kikuyu pasture with deep roots to extract nutrients from shallow ground water tables in the peel region. Kikuyu varieties will also be suited to pastures irrigated with solar power.

In our studies of marbling and low melting temperature beef it has become obvious that feeding plays a critical role. We will start feed comparison trials in 2017, with paired animals to reduce the effects of genetic variation. Future feed and pasture trials will be facilitated by using embryo splitting to create twins and eliminate genetic variability from future feed trials.

Vice-Chairman's Report

Mr. Anthony Lloyd

Much of the character of modern life depends on past scientific research. When we consider our current challenges we naturally turn to scientific endeavours to help find answers.

The Peel Region has many natural advantages, and quite a few challenges. The Foundation's decision to relocate to the region provides an opportunity to play a part in meeting those challenges through its research activities.

In 2016 the Western Australian government announced its "Transform Peel" initiative. At its heart is a focus on human food and the opportunity for agriculture in a peri-urban environment to support the economic development of the region.

During 2016 and continuing into 2017, the Foundation participated in a working group aimed at creating a facility for scientific research to support the food and agriculture industry in the region.

This is not a "top-down" approach. Farmers and others in the food and agricultural sector well understand the problems. The challenge is to foster "translational research" which will apply scientific knowledge to practical problems. It requires the involvement of practitioners in the field working closely with researchers in a multi-disciplinary facility. Part of the vision is to support new science and agriculture based business ventures.

The work of the planning group is well advanced. Land for the facility has been identified adjacent to North Dandalup townsite. The views of potential users of the facility have been canvassed and taken into account in facility planning. Town and traffic planning authorities have been consulted. Consultants have been engaged to prepare a detailed site plan and preliminary costings suitable for a development application to the local authority along with an independent comparative site analysis.

While the Foundation looks forward to the construction of a new facility, perhaps as early as 2018, we do not intend to be limited by that timetable. We expect to have available interim premises available during 2017 to house our ongoing activities within the Peel Region.

Treasurer's Report

Mr. Wayne Cox

2016 provided another cash flow surplus, resulting in $105,735 after non-cash addbacks. Continued rationalisation of our largest assets, the incubators, saw us maintain our level of lease income close to $200,000 during a tough rental market, marginally higher than in 2015.

This year we sold down a further 1,000,000 shares in GTG, after disappointing results from GTG and to focus on conserving and building our cash reserves for our pending relocation to North Dandalup during the 2017 year.

In the meantime, we are marketing the incubators for sale to convert more of our property into liquid assets.

Our cash reserves at 31 December 2016 are $1,245,000 in term deposit and $42,000 cash at the bank, up by $114,000 on the prior year.

During the year, the foundation has continued to keep costs at a minimum, with our major costs being wages, dropping from $106,495 in 2015 to $92,887 in 2016. We have however continued to spend funds on our Patent costs protecting our biggest asset, our Intellectual Property.

Wayne Cox
Treasurer

Honorary Secretary's Report

Mr Robert Giura

A Dispatch from the Peel Region

Robotic Dairy and Beef Farmer located at Keysbrook

I was invited to join the Board by the Chairman after having attended several CYO Foundation's scientific retreats where I gained an understanding about the Foundation's research into healthy beef with low melting temperature fat. This was of great interest to me because of my longstanding passion for producing high quality beef using Wagyu hybrids. The Foundation's more recent research for identifying and developing superior varieties of Kikuyu is a project that I am keen to be involved in as it could potentially provide significant productivity and environmental benefits, within the Peel region and throughout pasture based livestock production regions in Australia. This project has recently received considerable funding for research from the Peel Development Commission due to the efforts of our Vice-Chairman, Mr Tony Lloyd.

The weekly research meetings held at the 'Kielman Dairy' laboratory have provided regular updates on research and presentations by guest speakers. Regular attendees have comprised of representatives from the Shire of Murray and the Peel Development Commission. Some of these sessions, which included tours of operations and production facilities, have been hosted by local businesses; Keysbrook Leucoxene, C-Wise and Greenacres Turf Farm. A recent highlight was a highly informative presentation on Rumination by Dr. Reb. Other speakers of note include Hayley Norman from CSIRO, Rob Trengove, from Murdoch University and Kevin Hepworth from the Department of Agriculture and Food, Western Australia (DAFWA).

Deputy Research Director, Dr. Sally Lloyd, deserves a special mention for securing regular guest speakers as well as organising and convening the weekly sessions

Foundation Member Profiles and Reports

Obituary: Maxwell George MacDonald

Loving husband of Annette for 35 years and loving father of Colin, Deanne and Janise and Grandfather of 5 – boys passed away on March 21 2016.

Max was a long-term supporter and friend of the CY O'Connor ERADE Village Foundation having served as Vice Patron for more than 15 years.

Professionally Max has had a deeply satisfying and really full life. He started his working life at Crown Law after school as a junior clerk then built his way to become clerk of courts. He went then to the housing commission in administration which lead to his attending night school where he studied valuation to become a licensed valuer and he obtained certificate #32 – of which he was very proud, meaning he was one of WA's earliest valuers. Throughout his life, Max worked at bettering his situation whether through extra study, extra jobs. Max was on the board of Clubs WA amongst others. He was in the Public Trustee for many years when he often helped not just family and friends but people in need sort out probate and wills. He was a major figure in WA's land development industry through his role leading Taylor Woodrow Australia, which developed many thousands of residential home-sites throughout Perth and the regions. He was an inaugural councillor with UDIA WA (Urban Development Institute of Australia WA) and became a valued president and life member in WA, a National President and later a National Life member all before ending up as managing director of Taylor Woodrow. His friend Bill Griffiths said the thing about him was his absolute fairness, his solid integrity with both Government and the private sector. He also said he was a great 'giver'.
A long-term member of the Australian Army Reserve which commenced when he was called up at 18 for national service. The family is unsure if or when he resigned but to some extent he was involved with it right up until his passing.
Max had spent a fair amount of time in retirement in Ledge Point a he loved. Enjoying the ocean and either fishing, pulling his cray pots or just down at the beach chatting with eve-

ryone. He became a big part of the community, was a founding member of the GDEL, a company established to design, finance and build the Ledge Point Holiday Caravan Park. He was also a past president of the Ledge Point Community Council responsible for building the Ledge Point Golf Club and clubhouse.

Jason Wallis

Dr. Ted Steele

Ted Steele is a cellular and molecular immunologist of some 40 years standing who re-joined the CYO Foundation in 2014 after a 3-year absence. His main work in recent years has been on the mechanism of somatic hypermutation of antibody variable region genes and on rapid evolutionary mechanisms. The main theme running through all this work has been the role of RNA intermediates and reverse transcription in the generation of genetic diversity. Recently this has led to the analysis of molecular mechanisms in the early stages of cancer development. He has a deep fascination with the whole topic of ancestral haplotypes and hopes to pursue his interests around this theme at CYO.

Dr. H. "Reb" Rebhan, (DVM)

Currently working as a Veterinary Officer for the Australian Government Department of Agriculture and Water Resources.

Production Animal Medicine Registrar for Murdoch University's Veterinary School from 2008 to 2014. Before that, pet and production animal private practice veterinarian. 1984 Graduate University of Minnesota College of Veterinary Medicine. Author of All Things Strange and Wonderful. My Adventure as a Vet in Africa, a book everyone should read.

Joined due to the dogged persistence of Professor Roger Dawkins ongoing requests. I bring the insight from over 30 years as a production animal veterinarian- for whatever that is worth. I get to be one of the first people to see the exciting research outcomes from the foundation.

Mr Ben Dawkins **Deputy Honorary Secretary**

Innovation has become Prime Minister Turnbull's favourite topic. Although now a political buzz word, it is what happens on the ground that matters. The CYOEV Foundation has produced tangible results within Piara Waters with a smart combination of successful research and incubation of start-ups within in the business incubator buildings. The recent patents on melting temperature measurement and genetic markers for meat quality provide inspiration for all of us who support the Foundation. The prospect of another focused program in North Dandalup inclusive of business incubators means research and incubation of start-ups can again be combined to great effect.

The commercialization of the incubator buildings in Piara Waters is almost complete, the buildings will soon be subdivided and serviced to enable individual sale and a solid commercial return for the Foundation. As a board members, we are committed to ensuring that maximum value is extracted from these assets to enable reinvestment in the North Dandalup hub. Working with tenants, contractors and the authorities to prepare the buildings for sale is time consuming but essential work for the Board.

During my time on the Board I have had the opportunity to visit the business incubators at Curtin University, Fremantle Prison and Gosnells (the Agonis Centre). With the introduction of the North Dandalup site the Foundation is ideally positioned to provide incubation, development and commercialization services that will compare very favourably with other business incubators. The driver of this will be a renewed focus on agriculture and food production and the involvement of farmers in the Peel area. Real innovation, tangible results with commercial benefits. I expect Prime Minister Turnbull will soon be keen to visit the new hub in North Dandalup.

Ben holds bachelor degrees in Economics and Law as well as a Graduate Diploma in Business (Human Resource Management). Ben has worked in large corporates such as BHP Billiton as a Senior Human Resources Advisor and as a lawyer for several private law firms.

Mr Jason Wallis

C Y O'Connor Incubators Piara Waters

During the course of 2016 work continued on the proposal to release value from the Incubator Buildings on Erade Drive in Piara Waters. The proposal will see the five incubator buildings proceed through a subdivision process with a view to having each of the buildings ultimately subdivided on a separate green title allotment.

Once subdivision has been achieved, the Board may consider sale of the individual buildings and land to a third party. The buildings are likely to appeal to small business owners and investors given the flexible zoning and use allowances. The area of Piara Waters is now quite a mature catchment of some 3,000 homes, several primary schools and significant retail development. Consultants have provided an assessment of the underlying improved value in the order of $775,000 - $850,000 per lot. A comprehensive market comparison has not yet been undertaken however we feel this is broadly commensurate with unimproved commercial land values.

The five proposed incubator lots are now the subject of a valid WAPC conditional subdivision approval (WAPC 151813). Engineering approvals are currently being sought with the view to undertaking works mid 2017 in order to create titles circa September 17. Appointed engineers have undertaken an assessment of the peak cash requirement for the project which is in line with expectations.

Financial Statements

IMMUNOGENETICS RESEARCH FOUNDATION WA INC

T/A

THE C.Y. O'CONNOR ERADE VILLAGE FOUNDATION

67 161 830 965

FINANCIAL REPORT

FOR THE YEAR ENDED

31st December 2016

PASCOE PARTNERS ACCOUNTANTS

CERTIFIED PRACTISING ACCOUNTANTS

Liability limited by a scheme approved under

Professional Standards Legislation

IMMUNOGENETICS RESEARCH FOUNDATION WA INC
T/A
THE C.Y. O'CONNOR ERADE VILLAGE FOUNDATION
67 161 830 965

CONTENTS

IMMUNOGENETICS RESEARCH FOUNDATION WA INC
T/A
THE C.Y. O'CONNOR ERADE VILLAGE FOUNDATION
67 161 830 965
C Y O'CONNOR HI-TECH INCUBATOR CENTRE
DEPARTMENTAL TRADING, INCOME AND EXPENDITURE
FOR THE YEAR ENDED 31 DECEMBER 2016

	Note	2016 $	2015 $
C Y O'CONNOR HI-TECH INCUBATOR CENTRE			
OTHER INCOME			
Lease Rentals		198,980	198,706
EXPENSES			
Bad Debts Written Off	2	812	-
Depreciation - Plant & Equipment	3	2,462	2,928
Depreciation – Buildings		26,647	26,233
Incubator Management Fees		15,515	22,575
Repairs & Maintenance	4	7,597	12,456
		53,033	64,192
SURPLUS (DEFICIT) FROM INCOME AND EXPENDITURE		145,947	134,514
Opening Balance 1st January 2016		1,335,168	1,200,654
Closing Balance 31st December 2016		1,481,115	1,335,168

IMMUNOGENETICS RESEARCH FOUNDATION WA INC
T/A
THE C.Y. O'CONNOR ERADE VILLAGE FOUNDATION
67 161 830 965
RESEARCH & GENERAL FUND
DEPARTMENTAL TRADING, INCOME AND EXPENDITURE
FOR THE YEAR ENDED 31 DECEMBER 2016

	Note	2016 $	2015 $
RESEARCH & GENERAL			
Other Income		-	50,000
Donations/Bequests		167,494	179,943
Fund Raising		-	22,275
Meeting Registration Fees		7,345	9,140
Membership Fees		3,866	10,736
Laundry & Gaming Machine sales		715	1,610
		179,420	273,704
OTHER INCOME			
Interest Received		25,913	30,091
Reimbursements & Recoups		59	2,404
Profit / (Loss) on Sale of Non-current Assets		(9,030)	206,242
		16,942	238,737
EXPENSES			
Accountancy & Audit Fees		4,500	4,200
Advertising		540	2,252
Bank Charges		88	308
Cleaning		2,145	1,564
Consultancy Fees		8,198	2,750
Consultancy Fees - Jose Valenzuela		22,303	
Computer Expense		28	7,143
Depreciation - Plant & Equipment	3	18,213	18,587
Bad Debts Written Off	2	5,530	-
Fees & Permits		1,322	324
Freight & Cartage		279	75
Fund Raising Costs		-	5,022
Innovation Award		10,000	30,000
Insurance		10,526	10,227
Legal Costs		-	5,823
Meeting / Visitors / Honorarium		18,447	16,254
Consumables & Lab Supplies		9,829	23,143
Motor Vehicle Expense		756	3,147
Patent Fees		58,929	62,181
Printing & Stationery		715	1,117
Repairs & Maintenance	4	-	-
Subscriptions		-	313
Superannuation Contributions		8,157	11,763
Telephone		4,075	6,415
Travelling Expenses		5,907	-

IMMUNOGENETICS RESEARCH FOUNDATION WA INC
T/A
THE C.Y. O'CONNOR ERADE VILLAGE FOUNDATION
67 161 830 965
RESEARCH & GENERAL FUND
DEPARTMENTAL TRADING, INCOME AND EXPENDITURE
FOR THE YEAR ENDED 31 DECEMBER 2016

	Note	2016 $	2015 $
Unrealised loss / (gain) on Shares		168,233	(195,250)
Wages		92,887	106,495
Waste Disposal		522	934
		452,129	124,787
SURPLUS (DEFICIT) FROM INCOME AND EXPENDITURE		(255,767)	387,654
Opening Balance 1st January 2016		1,369,744	954,311
Transfer from Emergency Fund		-	5,000
Transfer from 7th MHC Evolution Meeting Fund		-	22,779
-		-	27,779
Closing Balance 31st December 2016		1,113,977	1,369,744

IMMUNOGENETICS RESEARCH FOUNDATION WA INC
T/A
THE C.Y. O'CONNOR ERADE VILLAGE FOUNDATION
67 161 830 965
BALANCE SHEET
AS AT 31 DECEMBER 2016

	Note	2016 $	2015 $
TRUST FUNDS			
RESERVES	5	2,595,092	2,704,913
TOTAL TRUST FUNDS		2,595,092	2,704,913
Represented by:			
CURRENT ASSETS			
Cash on Hand		1	1
Petty Cash		111	82
Deposits on Building Contracts		-	-
Westpac Max 112867		25,650	16,068
Westpac Research Account 10-4752		(2,694)	3,174
Westpac Community Solutions Account 1645		19,471	23,140
Term Deposits	8	1,245,336	1,131,024
Accounts Receivable		1,310	22,316
Loans - Unsecured	6	10,000	10,000
GST Refundable		10,393	11,165
TOTAL CURRENT ASSETS		1,309,578	1,216,970
NON CURRENT ASSETS			
Fixed Assets			
Computer Equipment		364,971	364,971
Less: Accumulated Depreciation		(363,834)	(363,117)
		1,137	1,854
Motor Vehicles		13,500	13,500
Less: Accumulated Depreciation		(12,781)	(12,572)
		719	928
Office Furniture & Equipment		11,612	11,612
Less: Accumulated Depreciation		(11,076)	(10,942)
		536	670
Laboratory Equipment		552,725	507,296
Less: Accumulated Depreciation		(440,977)	(423,824)
		111,748	83,472
Incubators		55,124	55,124
Less: Accumulated Depreciation		(43,554)	(41,092)
		11,570	14,032
Incubator Buildings		1,287,625	1,287,296
Less: Accumulated Depreciation		(287,411)	(260,764)
		1,000,214	1,026,532
Total Fixed Assets		1,125,924	1,127,488
Investments			
Listed Shares in Public Companies	7	132,183	325,417
Patents		54,750	54,750
Total Investments		186,933	380,167

IMMUNOGENETICS RESEARCH FOUNDATION WA INC

T/A

THE C.Y. O'CONNOR ERADE VILLAGE FOUNDATION

67 161 830 965

BALANCE SHEET

AS AT 31 DECEMBER 2016

	Note	2016 $	2015 $
TOTAL NON CURRENT ASSETS		1,312,857	1,507,655
TOTAL ASSETS		2,622,435	2,724,625
CURRENT LIABILITIES			
Trade Creditors		12,208	1,575
Incubator Bond Monies	9	7,970	11,490
Amounts Withheld	10	7,165	6,647
TOTAL CURRENT LIABILITIES		27,343	19,712
TOTAL LIABILITIES		27,343	19,712
NET ASSETS		2,595,092	2,704,913

IMMUNOGENETICS RESEARCH FOUNDATION WA INC

T/A

THE C.Y. O'CONNOR ERADE VILLAGE FOUNDATION

67 161 830 965

NOTES TO THE FINANCIAL STATEMENTS

FOR THE YEAR ENDED 31 DECEMBER 2016

1 Statement of Significant Accounting Policies

The director of the trustee company has prepared the financial statements of the trust on the basis that the trust is a non-reporting entity because there are no users dependent on general purpose financial statements. The financial statements are therefore special purpose financial statements that have been prepared in order to meet the needs of beneficiaries.

The financial statements have been prepared in accordance with the significant accounting policies disclosed below which the director has determined are appropriate to meet the needs of members. Such accounting policies are consistent with the previous period unless stated otherwise.

The financial statements have been prepared on an accruals basis and are based on historical costs unless otherwise stated in the notes. The accounting policies that have been adopted in the preparation of these statements are as follows:

Property, Plant and Equipment

Property, plant and equipment are carried at cost, independent or director's valuation. All assets, excluding freehold land and buildings are depreciated over the useful lives of the assets to the trust.

Increases in the carrying amount arising on revaluation of land and buildings are credited to a revaluation reserve in shareholders' equity. Decreases that offset previous increases of the same asset are charged against fair value reserves directly in equity; all other decreases are charged to the income statement. Each year the difference between depreciation based on the revalued carrying amount of the asset charged to the income statement and depreciation based on the asset's original cost is transferred from the revaluation reserve to retained earnings.

The carrying amount of plant and equipment is reviewed annually by directors to ensure it is not in excess of the recoverable amount from these assets. The recoverable amount is assessed on the basis of the expected net cash flows that will be received from the asset's employment and subsequent disposal. The expected net cash flows have not been discounted in determining recoverable amounts.

Property, Plant and Equipment (PPE)

Leasehold improvements and office equipment are carried at cost less, where applicable, any accumulated depreciation.

The depreciable amount of all PPE is depreciated over the useful lives of the assets to the trust commencing from the time the asset is held ready for use.

Leasehold improvements are amortised over the shorter of either the unexpired period of the lease or the estimated useful lives of the improvements.

IMMUNOGENETICS RESEARCH FOUNDATION WA INC
T/A
THE C.Y. O'CONNOR ERADE VILLAGE FOUNDATION
67 161 830 965
NOTES TO THE FINANCIAL STATEMENTS
FOR THE YEAR ENDED 31 DECEMBER 2016

		2016 $	2015 $
2	**Bad Debt Written Off**		
	Hi-Tech Incubator Centre	812	-
	Research & General Fund	5,530	-
		6,342	-
3	**Depreciation - Plant & Equipment**		
	Hi-Tech Incubator Centre	2,462	2,928
	Research & General Fund	18,213	18,587
		20,675	21,515
4	**Repairs & Maintenance**		
	Hi-Tech Incubator Centre	7,597	12,456
	Research & General Fund	-	-
		7,597	12,456
5	**RESERVES**		
	Reserves - Deficit	-	-
	7th MHC Evolution Meeting Fund	-	-
	C Y O'Connor Hi-Tech Incubator Centre	1,481,115	1,335,169
	Research & General Fund	1,113,977	1,369,744
		2,595,092	2,704,913
6	**Loans - Unsecured**		
	E J Steele	10,000	10,000
		10,000	10,000
7	**Listed Shares in Public Companies**		
	Shares in Listed Companies	9,012,500	9,762,500
	Less Provision for Diminution in Value	(8,880,317)	(9,437,083)
		132,183	325,417

Listed Shares in Public Companies include 12,016,667 shares held in Genetic Technologies Ltd. In 2004 the Foundation sold patents to Genetic Technologies Ltd. The agreement provided for the sale of the patents by the Foundation to Genetic Technologies Ltd for an amount of $15,000,000 and the granting of a licence by Genetic Technologies Ltd the foundation for the use of patents, at a cost of $2,500,000. The sale agreement provided the Foundation would receive net consideration of $12,500,000 to be Satisfied by 16,666,667 shares at an agreed price of 75 cents each. The shares were issued to the The Foundation on the 16th of June 2004. The shares were subject to escrow. Twenty percent of the total shareholding came out of escrow on each anniversary of the acquisition date. The ASX share price of Genetic Technologies Ltd at 31 December 2016 was 1.1 cents per share. Although the shareholding is considered a long term investment by the foundation, in accordance with normal accounting practice, the Foundation has provided $8,880,317 for the diminution in the value of the shares between the issue price and the market price at 31 December 2016.

IMMUNOGENETICS RESEARCH FOUNDATION WA INC
T/A
THE C.Y. O'CONNOR ERADE VILLAGE FOUNDATION
67 161 830 965
NOTES TO THE FINANCIAL STATEMENTS
FOR THE YEAR ENDED 31 DECEMBER 2016

		2016 $	2015 $
8	**Term Deposits**		
	Westpac Term Deposit 17-7599	100,000	125,000
	Westpac Term Deposit 18-6946	64,192	125,000
	Westpac Term Deposit 21-4134	128,071	125,000
	Westpac Term Deposit 21-4126	100,000	100,000
	Westpac Term Deposit 21-5671	125,000	125,000
	Westpac Term Deposit 21-9453	99,999	104,182
	Westpac Term Deposit 23-4151	101,685	100,000
	Westpac Term Deposit 23-4178	102,047	100,000
	Westpac Term Deposit 23-4186	102,557	100,000
	Westpac Term Deposit 21-1689	130,276	126,842
	Westpac Term Deposit 24-1511	27,317	-
	Westpac Term Deposit 24-4819	100,000	-
	Westpac Term Deposit 24-9548	64,192	-
		1,245,336	1,131,024
9	**Incubator Bond Monies**		
	Auslanka	1,120	1,120
	LimeLight	-	720
	Better Care Medical Centre	1,360	1,360
	Youme Clothing	450	450
	Spicy Wok	1,200	1,200
	Make Your Mark Art Cafe	1,040	1,040
	Infinity Wealth Solutions	-	-
	Renov8withus Pty Ltd	1,400	1,400
	Indebiti Solutions Pty Ltd	1,400	1,400
	Alight Scent	-	1,400
	Mr Stubby	-	1,400
		7,970	11,490
10	**Amounts Withheld**		
	Amounts Withheld from Salary, Wages and Other Payments	4,424	2,406
	Employee Superannuation Payable	2,741	4,241
		7,165	6,647

Compilation Report

COMPILATION REPORT
IMMUNOGENETICS RESEARCH FOUNDATION WA INC
T/A
THE C.Y. O'CONNOR ERADE VILLAGE FOUNDATION
67 161 830 965

We have compiled the accompanying special purpose financial statements of THE C.Y. O'CONNOR ERADE VILLAGE FOUNDATION which comprise the balance sheet as at 31 December 2016, profit and loss statement and cash flow statement for the year then ended, a summary of significant accounting policies and other explanatory notes.

The specific purpose for which these special purpose financial statements have been prepared is to provide information relating to the performance and financial position of the incorporated body that satisfies the information needs of the director.

The responsibility of the Director

The director is solely responsible for the information contained in the special purpose financial statements and has determined that the basis of accounting used is appropriate to meet their needs and for the purpose for which the financial statements were prepared.

Our responsibility

On the basis of information provided by the director, we have compiled the accompanying special purpose financial statements in accordance with the financial reporting framework described in Note 1 to the financial statements and APES 315: Compilation of Financial Information.

Our procedures use accounting expertise to collect, classify and summarise the financial information, which the director provided, into a financial report. Our procedures do not include verification or validation procedures. No audit or review has been performed and accordingly no assurance is expressed.

The special purpose financial statements were prepared exclusively for the director. We do not accept responsibility to any other person for the content of the special purpose financial statements.

Name of Firm: Pascoe Partners Accountants
Certified Practising Accountants

Name of Partner: Wayne Michael Cox
MTax, Dip FS(FP), CPA, CTA

Address: Level 2, 100 Havelock Street West Perth

Dated: 15.3.17

Auditor's Report

The C.Y. O'Connor Erade Village Foundation
Independent Auditor's Report
to the Members of
The C.Y. O'Connor Erade Village Foundation
Incorporating
The Immunogenetics Research Foundation
And
The Institute of Molecular Genetics and Immunology Incorporated

---

We have audited the accompanying financial report, being a special purpose financial report, of The C.Y. O'Connor Erade Village Foundation (the association), which comprises the committee's report, the assets and liabilities statement as at 31 December 2016, the income and expenditure statement for the year then ended, notes comprising a summary of significant accounting policies and other explanatory information, and the statement by members of the committee.

Committee's Responsibility for the Financial Report
The committee of The C.Y. O'Connor Erade Village Foundation is responsible for the preparation of the financial report, and has determined that the basis of preparation described in Note 1 is appropriate to meet the requirements of the Associations Incorporation Act (1987) and is appropriate to meet the needs of the members. The committee's responsibility also includes such internal control as the committee determines is necessary to enable the preparation of a financial report that is free from material misstatement, whether due to fraud or error.

Auditor's Responsibility
Our responsibility is to express an opinion on the financial report based on our audit. We have conducted our audit in accordance with Australian Auditing Standards. Those standards require that we comply with relevant ethical requirements relating to audit engagements and plan and perform the audit to obtain reasonable assurance whether the financial report is free from material misstatement.

An audit involves performing procedures to obtain audit evidence about the amounts and disclosures in the financial report. The procedures selected depend on the auditor's judgment, including the assessment of the risks of material misstatement of the financial report, whether due to fraud or error. In making those risk assessments, the auditor considers internal control relevant to the association's preparation of the financial report that gives a true and fair view, in order to design audit procedures that are appropriate in the circumstances, but not for the purpose of expressing an opinion on the effectiveness of the association's internal control. An Audit also includes evaluating the appropriateness of accounting policies used and the reasonableness of accounting estimates made by the committee, as well as evaluating the overall presentation of the financial report.

We believe that the audit evidence we have obtained is sufficient and appropriate to provide a basis for our audit opinion.

Qualification
In common with most charitable institutions receiving cash donations, the entity has determined that it is impractical to establish controls over cash donations before they are banked and recorded in the accounting system. As a result our audit was restricted to the amounts recorded in the accounting system.

The C.Y. O'Connor Erade Village Foundation
Independent Auditor's Report
to the Members of
The C.Y. O'Connor Erade Village Foundation
Incorporating
The Immunogenetics Research Foundation
and
The Institute of Molecular Genetics and Immunology Incorporated

---

Opinion

In our opinion, except for the effects on the financial report, if any that the above qualification may have on the financial report of The C.Y. O'Connor Erade Village Foundation, the report presents fairly, in all material respects, the financial position of The C.Y. O'Connor Erade Village Foundation as at 31 December 2016 and its financial performance for the year then ended in accordance with the accounting policies described in Note 1 to the financial statements, and the Associations Incorporation Act (1987).

Basis of Accounting and Restriction on Distribution

Without modifying our opinion, we draw attention to Note 1 to the financial report, which describes the basis of accounting. The financial report has been prepared to assist The C. Y. O'Connor Erade Village Foundation to meet the requirements of the Associations Incorporation Act (1987). As a result, the financial report may not be suitable for another purpose.

S. William

Steve Williams FIPA
Auditor

Dated this 22ND day of MARCH 2017

Auditors Independence Declaration
Under Section 307C of the Corporations Act 2001

To Members of the C.Y. O'Connor Erade Village Foundation

I declare that, to the best of my knowledge and belief, in relation to the audit of the C. Y. O'Connor Erade Village Foundation for the year ended 31 December 2016 there have been:

a) No contraventions to the auditor independence requirement of the Corporations Act 2001 in relation to the audit; and

b) No contraventions of any applicable code of professional conduct in relation to the audit.

S. William

Steve Williams FIPA
Auditor

Dated this..22ND........day of........MARCH....................2017

Articles

The Old Cottage Kathleen Room of the C.Y. O'Connor Tavern

The Old Cottage was built on the Livingston Estate on Nicholson Road near the present Canning Vale industrial park. In the early 1950s, it was moved here plank by plank, including the hand-sawn weatherboards visible on the front verandah.

The first occupant was the celebrated stockman, Boer War veteran and gambler, Bill Loutitt. According to legend, he is recognized in 'Such is Life' by "Tom Collins" and 'The Cattle King' by Ion Idriess. His yarns referred to the Durack drives, George and Lang Hancock, C Y O'Connor and family when they arrived in Albany, the Davis family in Karridale and many others.

During the next decade, the Dawkins boys, would ride their bicycles or ponies from Peppermint Grove to spend school holidays working on the farm with Bill and Jack Kelly. At that stage, the Old Cottage had only two rooms with a make-do kitchen containing a meter's camp stove and a Coolgardie safe. At the back verandah was the farm shed containing horse feed and fertilizer. Dugite snakes were everywhere leading us to test and empty the bed roll each night.

By the late 1960's there was a separate shed, a shower recess and an elementary septic tank. John Gill replaced Bill when he retired to Sunset in Dalkeith.

In 1970 relative major improvements were made including a rather modern bathroom and laundry, serviced by a septic tank next to the vine which still flourishes today. Electricity was provided by a cable from the generator at the Homestead on the top of

the hill. The back stables (now S Block) behind the pergola include a study where many medical publications and theses were written.

The Old Cottage hosted many gatherings including those associated with state and federal campaigns. These events were recorded in several short stories published by well-known local authors.

A further renovation was required when one of the adjacent pine trees fell across the entire cottage fortunately, somehow, avoiding the sleeping occupants.

During the 1980s and 90s there were a series of treasured occupants including Sue and Keith Mellican, Brian and Mavis Chambers and Peter and Ivana Tucacovic. The Old Cottage became the quarters for C Y O'Connor ERADE Village Foundation staff and later the office for the builders of the CYO Tavern.

The most recent renovation was completed in 2016. This included the restoration of the two front rooms to form the current "Kathleen Room" named for Kathleen O'Connor, aka the 'Artist in Exile' and the subject of a series of biographies and critiques listed below. Her photographic portrait dominates the room. Note also some prints of her paintings reflecting some of distinct Parisian, Fremantle and Perth periods. Her portrait of Bill Loutitt is in the corner where the pine tree fell.

Written by: E/Prof R L Dawkins whose great aunt was Kathleen O'Connor.

Select Bibliography:

1. PAE Hutchings and Julie Lewis, Kathleen O'Connor: Artist in Exile, Fremantle Arts Centre Press, Fremantle Western Australia, 1987.
2. Gooding, Janda, Chasing shadows: the art of Kathleen O'Connor, Craftsman House, East Roseville, NSW.
3. McCulloch, Alan, Encyclopaedia of Australian Art (Hutchinson of London 1968).

4. Hutchins, Patrick "Last Link with Impressionism", Bulletin, Sydney, 4 March 1967.
5. James Watson, 'O'Connor, Kathleen Laetitia (1876–1968)', Online: Australian National University, 2006.
6. Merab Tauman, The Chief, C Y O'Connor, University of Western Australia Press, 1978.
7. A G Evans, C Y O'Connor: His Life and Legacy, UWA Press, 2002.
8. O'Connor, Charles Yelverton (1843-1902), Australian Dictionary of Biography – Online edition.
9. Thomas Arthur Guy Hungerford, AM (5 May 1915 – 19 June 2011), popularly known as T. A. G. Hungerford, was an Australian writer, noted for his World War II novel The Ridge and the River, and his short stories that chronicle growing up in South Perth, Western Australia during the Great Depression.

Recent Foundation Publications

Analysis of Haplotype Sequences.

Lloyd SS, Steele EJ, Dawkins RL.
In: Kulski JK, Next Gener. Seq. - Adv. Appl. Challenges. InTech; 2016. p. 345–68.

Available from: http://www.intechopen.com/books/next-generation-sequencing-advances-applications-and-challenges/analysis-of-haplotype-sequences

Abstract
In this era of whole-genome, next-generation sequencing, it is important to have a clear understanding of the concept of "haplotype". We show here that most of the important regions of the genome can be described in terms of polymorphic frozen blocks (PFB). At each PFB, there are numerous, even hundreds, of alternative ancestral haplotypes. Haplotypes, not genes, can be regarded as the principal unit of inheritance.
We illustrate how sequence data can be analysed to reveal and define these ancestral haplotypes.
Keywords: Ancestral haplotypes, Polymorphic frozen blocks, Genomic evolution

Perspective on The Production and Transmission of Certain Eye Defects
Author: Edward J Steele PhD

In: Levin M, Adams DS. Inheritance. Ahead of the Curve.
IOP Publishing; 2016. p. 3–77.

Origin of Congenital Defects: Stable inheritance through the male line via maternal antibodies specific for eye lens antigens inducing autoimmune eye defects in developing rabbits in utero

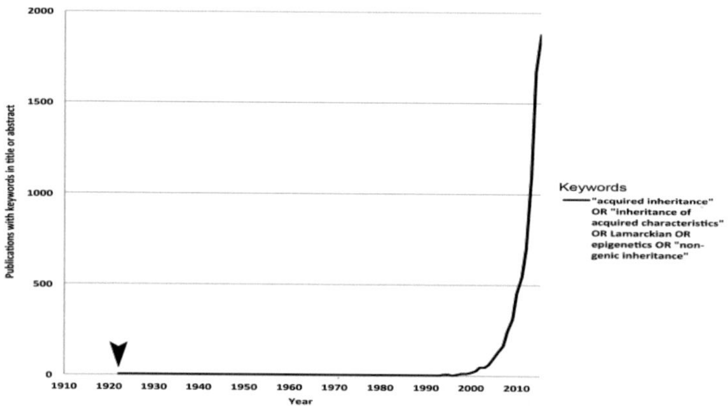

From Guyer (1922) The production and transmission of certain eye defects.

In April 1922 Michael F Guyer, Professor of Zoology at the University of Wisconsin, stood and addressed the International Congress of Othalmology in Washington D.C. He presented the results of a remarkable series of 'acquired inheritance' experiments which he and Elizabeth A Smith had been conducting at the Zoological Laboratory since 1916 (Guyer 1922). He also had with him live rabbits bearing defective eyes so that the gathered opthalmologists could examine some of the evidence for themselves. Guyer and Smith showed that specific antibodies for eye lens antigens crossed the placenta in rabbits producing offspring with severe eye defects. By itself this is perhaps not surprising. They then showed that these induced eye defects were passed on to future generations arising spontaneously in progeny rabbits via the male or female lines without any further immunization or antiserum treatment of subsequent parental rabbits with eye lens antibodies (Guyer and Smith 1918, 1920, 1924). In my view these transgenerational genetic data are on a historical par with Mendel's landmark publication on inheritance of genetic factors affecting characteristics of pea plants 60 years earlier.

...

Full Article available from:
http://dx.doi.org/10.1088/978-0-7503-1326-1ch3

In The Press

Tech group tags a new market

JOHN ROSS
SCIENCE WRITER

Bluechip managing director Andrew McLellan

News from *Torr Down Stud*

SUPER SOUTH DEVONS

Torr Down Marvel 3, a ten-year-old Homozygous Poll sire, bred and owned by John and Sue Harvey of Trentham, Victoria, Australia, has been sold to two prominent Red Wagyu Akaushi breeders. The new owners are Melaleuca Red Wagyu, Canning Vale, Western Australia and Mazda Wagyu International Pty Ltd Mossiface, East Gippsland.

The two studs wish to promote polled cattle in the Red Wagyu Akaushi breed and Torr Down Marvel 3 has this ability along with the attributes of faster growth, excellent carcass specifications especially eye muscle, together with his outstanding docility, plus the South Devon's renowned maternal traits.

Resnick urges farmers to unite, speak up

Continued from Page 19

A conversation" noted California's 2015 farm production of $US47bn was down sharply from the $US54bn the previous year. The expectations for the year ahead are only for a slight improvement.

"Our biggest vulnerability continues to be the global supply and demand picture for dairy — our number one ag commodity — and low prices. Overall, I think total ag value in 2016 could be slightly higher than 2015," Karen Ross, secretary of the California Department of Food and Agriculture, said in the paper.

At the LA forum Pratt challenged America's most populous state to double the value of its food exports to more than $US40bn annually as his company prepares to build its first hi-tech cardboard box factory in California.

California's challenges are not dissimilar to Australia's: water, labour, prices, costs and regulation.

On the first and arguably most pressing issue, the Australian Consul-General in California, Chelsey Martin, told the dinner that water management and

Stewart Resnick and Anthony Pratt at the Global Food Forum event in Los Angeles

The New York Times http://nyti.ms/2e4eha2

WELL | EAT

Should We Be Scared of Butter?

Personal Health

By JANE E. BRODY OCT. 31, 2016

Paul Rogers

Is butter, along with other sources of saturated fats, back on the table, as many have recently claimed?

It is. Just not in the quantities the meat, dairy and fast-food industries might have you ingest.

Unless you have a medical condition that dictates otherwise, there's no reason to cut anything – not butter, ice cream or Porterhouse steak — completely from your diet as long as you mainly eat plant-based foods (vegetables, fruits and whole grains), lean animal protein and fish and don't go overboard on foods rich in saturated fats that can cause harm in excess.

That's the conclusion of the best available evidence I've reviewed for maximizing the health of body and brain and enjoying a long life.

The unending controversies about a wholesome diet provide much fodder for this column. An extensively researched book by the science writer Nina Teicholz ("The Big Fat Surprise: Why Butter, Meat and Cheese Belong in a Healthy Diet"), published in 2014, has raised serious questions about the evidence that nearly 40 years ago prompted the Senate Selec Committee on Nutrition and Human Needs to recomme Americans follow a diet saturated fats and chol curb what was then epidemic of heart cardiovascular d

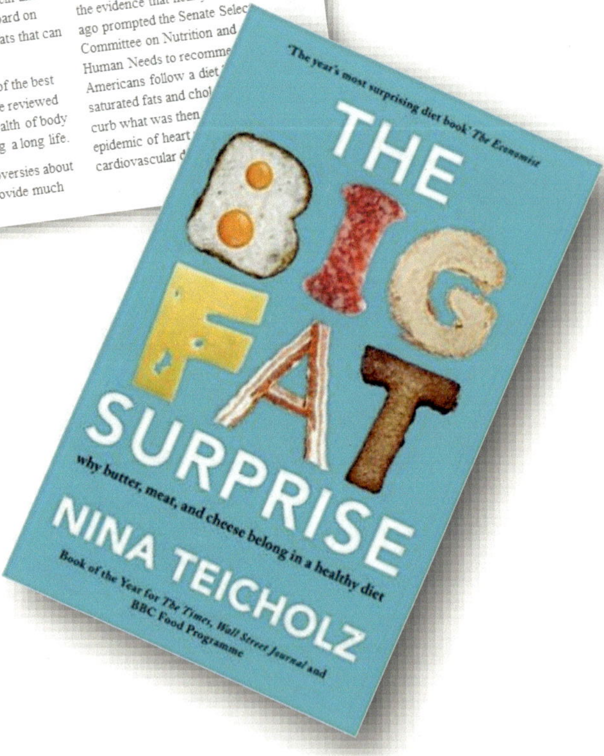

Travel Reports

Australian Wagyu Conference 2016

Plant Breeding Institute, University of Sydney

Sally Lloyd

On Wednesday 11th May I drove from Canberra to Cobbitty (30 minutes south west of Sydney) to visit Peter Sharp and his PhD student Duncan Fraser, to view their current experiments and to explore the possibilities of a collaboration with CYO. I also met their commercial partner George Hatton who runs a turf farm nearby.

Current Experiments

Their Kikuyu varieties were collected from various locations around the country. The aim of the collection was to get the maximum variety of origins and growing conditions. Varieties include one known to be first import 1920 origin and one variety growing in salt marsh. Established varieties, Whittet, Noonan, Village Green and Kenda are also used in experiments to provide a benchmark.

Peter's previous student Ben, identified genetic variation via RAPD and found three main groupings, related to country of origin in Africa. Ben also line bred within these varieties and F2 plants from these are now being used in the experiments.

Phenotypic variations: growth rate, fineness of leaf, maximum height, tendency to spread (tall forms and prostrate forms significantly different, greenness, maintain green through cold.

Plots of 9 different kikuyu varieties compared in irrigated and "drought" conditions – unirrigated plots get natural rainfall. Varieties of kikuyu were tested for salt tolerance in tubes and some grew in concentrations as high as sea water.

Measurements: measuring Amino acids likely to be related to salt and drought tolerance. Testing was via NMR. Other chemicals were also measured, but not analysed. Lignin and other factors in nutrient content and palatability may be available with this testing.

Current Commercialisation - Turf

One variety has been selected for use as a turf variety and they are currently going through the necessary tests for registering this variety to get Plant Breeder Rights.

Main features of this turf variety are: fine leaf and short inter-node distance. Remains short (less than 15cm) without mowing. Low matting (reduced weight of turf rolls)

Future Commercialisation – Pasture

Some suggestions from Hatton that the variety being registered could also be suitable for pasture, that the short stem length and greenness would correlate with palatability and that matting was also a problem for their local dairy farmers. My impression of this variety is that growth rate will be too low for a pasture plat and that the short leaf length will also be difficult for cows to eat.

Other varieties currently being tested for drought tolerance may have more promise for pasture. Dry matter or Metabolizable energy production rates have not been measured. Palatability and nutritional value has not been tested. The Cobbitty campus also has a dairy farm for teaching purposes and an analytical facility with a Gas Chromatography Mass Spectrometer that will test for a wide variety of chemical compounds.

Duncan identified very poor seed survival rates as a challenge for producing a pasture plant.

Potential for Collaboration

What does CYO bring to the collaboration?

- Collaborations involving multiple commercial partners and research organisations have access to grants not available to any single body.
- The ability to test in local conditions with deep sand and accessible ground water.
- Trial paddocks under irrigation and rotational grazing.

The conditions for drought tolerance at Cobbitty are not the same as in the Peel Region. Soil, ground water and climate are all different. Cobbitty has sandy soil, to a depth of 1m over a clay base. Summer rainfall was low this year but is still generally higher than Perth rainfall. Kikuyu plants in Autumn are starting to yellow due to cold.

Duncan has plans to measure root depth under controlled conditions by planting in tubes, but it is not a major factor for their region since they expect the kikuyu to be getting 90% of its nutrient and water from the first metre.

We should measure root depth by core samples under field conditions containing ground water at various depths and salinity.

Any collaborative agreement needs to consider ownership of IP. George Hatton has put $4 million into the project so far and is looking for commensurate returns. He expects to own the plant breeding rights for any variety coming out of their program, including those for pasture. CYO would gain little from testing to find which of their varieties would be best suit Beef and Dairy production in the Peel Region if they retain all Plant Breeding Rights.
We have expertise in DNA markers (BACD) and a point of difference to satisfy local conditions. We would benefit from access to the varieties they have already developed and tested, particularly with their superior expertise in plant breeding. We bring a greater knowledge of pasture and cattle requirements, although they have alternatives collaborators to access similar knowledge.
Peter Sharp was interested in establishing a collaboration but not willing to send samples of any sort until an agreement over the details of collaboration and IP sharing is worked out.

Scientific Retreats 2016

Weekly Research meetings

The foundation holds research meetings every Wednesday 1pm – 2pm where CYO researchers and visiting scientists present their latest work and producers and businesses can bring up local issues and observations. CYO members and friends are welcome to participate and lunch is provided.

Throughout the past year the CY O'Connor foundation has provided an incubation space for my seed business "Grubs Up" cricket farms.

The foundation has provided a professional platform in which I have been privileged enough to engage with various academics, and business professionals in order to push my knowledge and professional development for future business success.

CY O'Connor foundation has actively supported my business endeavours, through mentoring, retreats, and consistently encouraging my business to apply for many capacity building and funding opportunities.

Paula Pownall

Regular participants in 2016:

Brett Flugge, Dean Unsworth - Shire of Murray
Amy Denholm, Gary Watson, Justin Wolfgang, Orna Tippett, Jessica Bell, Michaela Tibi C-Wise
Tony Lloyd, Roger Dawkins, Sally Lloyd, Ted SteeleCYO Foundation
Reb Rebhan WA Department of Agriculture
Hillary Nath Eco Advise
Adrian Pitsikas Green Acres Turf Farm
Paula Pownall Grubs Up
Ray Candy Irwin Hunter
John Dawkins, Lindsay Baker Jose Valenzuela, Kelly Manning Melaleuka Stud
Louise Smyth Notre Dame University
Janine Hatch, David Arkwright, Linda Cunningham, Andrew Ward Peel Development Commission

Alan Peggs	Pardoo Station
Geoff Manning	Shade River Dairy
Bruce Campbell	
Graeme Brickwood	Serpentine Transport
Rob Giura	Mucca Dairies
Peter Bates	

Visiting Presentations

Keysbrook project Overview of Operations
Marc Morris/ Shaun O'Neill, MZI
Professor Ray Willis — Future Smart strategies
Nutrient Management for Wagyu Steers — Yoichi Sakaguchi
The function of soil biology to predict soil performance — L Abott
Improving Kikuyu with new Legumes — John Howieson
Livestock Disease Investigation Subsidies: Why and How
Kevin Hepworth, WA Department of Agriculture
Darren Wallis — Land Corp

Thanks to C-Wise and Green Acres Turf for hosting some sessions.

Selected Talk:

Dr. Rebhan

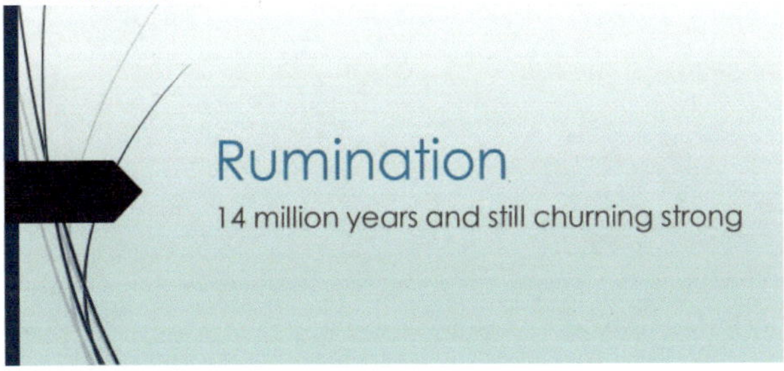

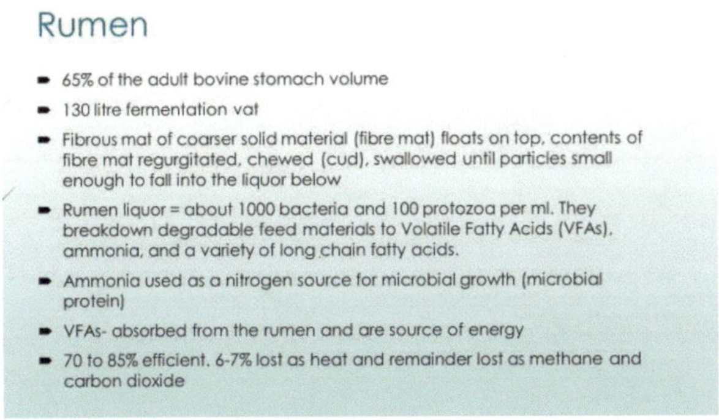

God's design

Forage (cellulose and hemi-cellulose) fermented into VFA's + microbial protein

Beauty in its simplicity and it doesn't get much simpler.

What we do

Forage (cellulose and hemi-cellulose) + Concentrates (grain, protein rich meal) + Fat + Bypass fat (Ca-lipids = soap) + Bypass protein (animal sources) + Non Protein nitrogen (urea) + Vitamins + Minerals + Buffers + Alkalizing agents + Yeast + Ionophores + Virginiamycin + Mycotoxin binding agents ferment to

VFA's + microbial protein + Bypass abomasal digested fats and protein.

(How could anything go wrong?)

Why?

- Human marathon runners- 2.6 x BMR
- Polar sled dogs- 3.4 x BMR
- Dairy cow at peak lactation- 5.3 x BMR WOW!

  The Energy Bank Charges 16%

Protein and rumen fermentation

- Rumen degradation/fermentation. Protein – amino acids-deamination (ammonia)- rebuild as microbial protein (Inefficient)
- True digestion- protein reaches abomasum (true stomach) digested into amino acids and absorbed (Efficient).

Accomplished with bypass protein = animal sources. Blood meal (80), meat and bone meal (75) (BSE) and fish meal (70). Plant sources – cotton seed, linseed meals, canola and soy low (30- 35).

- Non-protein nitrogen (Urea). NPN + starch = microbial protein. (Cost efficient but limited and Highly TOXIC)

Energy- Volatile Fatty Acids

- Rumen Fermentation- Volatile Fatty Acids (VFA)= 70% energy supply.
- VFAs absorbed through rumen epithelium. Epithelial surface area expanded by well-vascularized papillae. From the rumen to the liver and then beyond.

Acetic acid absorbed unchanged. It is utilized minimally in the liver and is oxidized throughout most of the body to generate ATP. Another important use of acetate is as the major source of acetyl CoA for synthesis of lipids (milk fat, SQ fat, HKP fat).

Propionic acid is also absorbed unchanged. It is almost completely removed from portal blood by the liver. Within the liver, propionate serves as a major substrate for gluconeogenesis, which is absolutely critical to the ruminant because almost no glucose reaches the small intestine for absorption. Milk sugar, lactose = galactose + glucose. **Glucose needed for marbling fat.**

Butyric acid is metabolized by the rumen epithelium to beta-hydroxybutric acid, a type of ketone body. Beta-hydroxybutyric acid is oxidized in many tissues for energy production. Brain and mammary tissue can not use as an energy source. Dairy cow ketosis- milk drops and brains swim.

Lipids

- Triglycerides- major lipid type found in cereal grains, oilseeds, animal fats
- Glycolipids- major lipid type found in forages

Breakdown of dietary lipids is rapid. Glycerol and sugars released from triglycerides and glycolipids are fermented into VFA.

Fatty acids released in the rumen pass to the abomasum and small intestine, primary site of absorption. Not absorbed through rumen wall.

Unsaturated fatty acids are toxic to many species of beneficial rumen bacteria.

Due to anaerobic environment in the rumen there is an excess of hydrogen that the microbial population is continually interested in getting rid of (why methane is produced). Unsaturated fatty acids are quickly hydrogenated to saturated fatty acids.

Ruminants-oxidative or fuel use of long-chain fatty acids is more limited than most non-ruminants. Acetate (VFA) the abundant oxidative fuel. For cows in negative energy balance (peak lactation), **only substantial use for dietary fat is milk fat synthesis.**

Added fat has no place in a beef finishing ration, unless it is -30 C, **feed for propionate**

Lets Talk Fermentation

- Normal Fermentation predominantly Fibre diet

93 acetate + 20 propionate + 5 butyrate + 61 carbon dioxide + 37 methane + 25 water (Propionate 16% VFA pool)

- Fermentation added concentrates (starch/grains)

61 acetate + 33 propionate + 13 butyrate + 59 carbon dioxide + 32 methane + 25 water (Propionate 31% VFA pool, 8% less carbon lost)

- Fermentation added concentrates + ionophores

50 acetate + 49 propionate + 13 butyrate + 46 carbon dioxide + 19 methane + 25 water (Propionate 44% VFA pool, 35% less carbon lost)

How do ionophores do that?

- Ionophores-decrease rumen gram positive bacteria in relation to gram negative.
- Gram-positive bacteria produce less desirable metabolites- acetate (we want propionate), hydrogen (acid), methane, ammonia, and lactic acid.
- Inhibit gram-positive hyper ammonia producers. Less degradation of protein in rumen. More protein reaches abomasum = true digestions =more efficient.
- Inhibit gram positive lactic acid producers. Less chance of rumen acidosis.
- Not all gram positive bacteria are bad bacteria but all bad bacteria are gram positive

Ionophores and Milk

Figure 1. Percentage changes in various responses due to feeding monensin to dairy cows across 77 trials (Duffield et al., 2008)

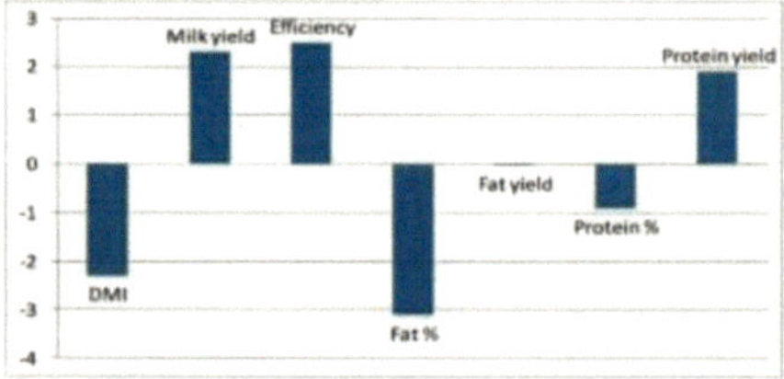

Ionophores & Where's the Beef?

- Feed efficiency – 17% improvement
- Average Daily Gain- 5% improvement
- Feed consumption- 13% decrease
- Marbling enhanced

4 week old calf fed milk and grain

4 week old calf fed milk and grain

6 week old calf fed milk and grain

8 week old calf fed milk and grain

12 weeks of age- Diet Milk and hay

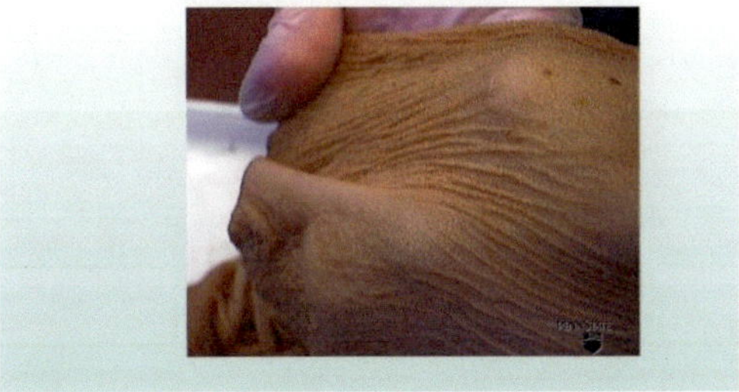

Rumen development

- Birth- calf's rumen the size of a softball
- 8 weeks- size of a plastic grocery bag large enough to carry two, 2-L bottles
- From 8-24 weeks- rumen double five times.
- Papillae growth directly correlated with diet. Feeds high in starch rapidly ferment and yield higher amounts of the VFA butyrate. Butyrate enhances rumen papilla growth-**What we want.**
- Fibre rich diets, while helpful in producing saliva flow and strengthening rumen wall muscles, impede rumen development and slow calf growth.

Take home message- get the hay out of the dairy calf pens! Creep up!

Creep Feeders for Beef

Do Finishing Beef Cattle Need Forage?
Da Man Proved No
50 Years later and still going strong!
Tend-R-Leen®
Doughboy
The original no roughage dairy beef feeding program
MOLASSIE FEED
Doboy®

What does pushing propionate accomplish?

- Carcass characteristics of Holstein steers fed whole corn and concentrate ration with or without access to long hay.

	Tend R Leen No Roughage	Hay (limited), Whole Shelled Corn, Concentrate	Free Choice Hay, W shell corn, Concentrate
Carcass wt. (lbs)	684	658	690
Dressing Percentage	59.4%	58.6%	60.1%
Marbling Score	5.7	5.2	4.9
Avg. Quality Score	Choice	Choice -	Select +
% Grading low choice or >	72.7%	50.0%	33.3%
Ave age slaughter (months)	14	16	19

What does pushing propionate accomplish?

- Carcass characteristics of Holstein steers fed whole corn and concentrate ration with or without access to long hay.

	Tend R Leen No Roughage	Hay (limited), Whole Shelled Corn, Concentrate	Free Choice Hay, W shell corn, Concentrate
Carcass wt. (lbs)	684	658	690
Dressing Percentage	59.4%	58.6%	60.1%
Marbling Score	5.7	5.2	4.9
Avg. Quality Score	Choice	Choice -	Select +
% Grading low choice or >	72.7%	50.0%	33.3%
Ave age slaughter (months)	14	16	19

What do you need to do now?

- **Get the bloody hay out of the dairy calf pens! Fresh water available at all times. Feed a grain based calf starter to promote rumen development.**
- **Creep Feed beef calves. Promote rumen development.**
  > Weaning wt
- **Beef finishers- do everything in your power to enhance propionate production. > Marbling, > ADG, < Time on Feed**

14th-15th April: Problem-Solving in Food and Agriculture

Thursday 14th April:

Function Room, CYO'Connor Pub &Tavern

Introduction

CYO Healthy Beef Research Program & Ancestral Haplotypes
Human c17 Synteny and Regulation of SREB-FASN Bovine c19 Haplotypes — *Ted Steele/Roger Dawkins*

Symposium 1 The CYO Pasture & Kikuyu Research Program

Directions for Kikuyu Research — *Tony Lloyd*
Research on Kikuyu, Plant Breeding Institute, University of Sydney — *Peter Sharp*
Seed Quality Pasture Grasses — *Ray Candy*
Using Gene Silencing to Control Plant Pests, Murdoch University — *Mike Jones*
Kikuyu Quality and Dairy Industry — *Rob Giura*
Plant Breeding Rights & Nutrient Stripping — *Adrian Pitsikas*
BACD-PCR Marker patterns and Kikuyu Varieties — *Sally Lloyd*

Symposium 2 Problem-Solving Group of CYOEVF
Convenor Roger Stubbs

Planning session with the intention of defining the Peel group and its facilities.

The Overall Plan — *Roger Stubbs*
Funding assistance — *David Arkwright*
Infrastructure assistance — *Dean Unsworth/Brett Flugge*
Surface waters and run off — *Anthony Lloyd*
Artesian water use — *Rob Giura*
Dairying in the region — *Geoff Manning*
Livestock — *Peter Bates/ Bruce Campbell*
Yangedi Hamlet (Housing Development) — *Jason Wallis*

Participants: Roger Stubbs – Convenor; Tony Lloyd; David Arlwright, Peel Regional Grants Dean Unsworth – Murray shire; Brett Flugge – Murray shire; Graeme Brickwood –Serpentine Transport; Jonathan Emanuel – Emanuel Brothers; Peter Bates- Peel Farmer perspective; Rob Giura – Peel Dairy; Bruce Campbell – Peel Livestock Farmer perspective; Geoff Manning – Peel Dairy perspective; Peter Bates – Peel livestock

Topics: Funding; Surface waters and run off; Artesian waters; Soil quality – livestock, dairy, grains; Dairy; Pastures; Feed lots; Quality Livestock

Symposium 3: Livestock Selection for Quality Beef

CYO Healthy Beef Program *Roger Dawkins*
Recent Advances in Muscular Dystrophy Research *Frank Mastaglia*
Wagyu Crossbreds on Pardoo in the Kimberley *Alan Peggs/Jose Valenzuela*
Wagyu Genetics and Low Tm Intramuscular Fat *Sally Lloyd*
The WA Wagyu Co-op *Peter Gilmour*
The Polling of Red Wagyu using the Senepol Breed *Barbara Roberts-Thomson*
Beef Cattle in North West China - Opportunities for Wagyu *Dom Bayard*
Wagyu and F1 –How Many Days on Feed? *Sally Lloyd*
Progress with use of Sexed Semen *Dom Bayard*
Eye Cancer in Cattle and Potential Cure *Martin Ashdown*
Myostatin Secretion as a Player in Low Tm Beef *Ted Steele*

CYO Foundation Annual General Meeting

Conference Dinner

CYO Pub & Tavern Function/Conference Room
Speakers:
Food Security, *Herbert Rebhan*
Stories on Wagyu and Other Breeds,
Barbara Roberts-Thomson, Frank Mastaglia, Peter Gilmour, Rob Giura.

Friday 15th April

CYO Pub & Tavern Conference Room

Symposium 4: Consumers and the Environment in Agriculture

The CYO Pub Paddock-to-Plate Program *John Dawkins*
Wagyu Crossbreds on Pardoo in the Kimberley *Alan Peggs/Jose Valenzuela*
What Australian Wagyu Farmers Looks for in Wagyu Production *Dom Bayard*

Symposium 5: Student Sessions

Agribusiness Degree at Charles Sturt University *Christine Storer/Wendy Dymond*
Charles Sturt Agribusiness Student Presentations *Various students*
Lebeckia: turning sand into summer feed *Tom Edwards*

Symposium 6: Next Generation Sequencing & Ancestral Haplotypes

Next Generation Sequencing (NGS) — *Yurek Kulski*
Review of NGS re Livestock Genetics — *Roger Dawkins*
Ancestral Haplotypes as Detected by NGS — *Sally Lloyd*
Ancestral Haplotypes as an Analytical Tool. e.g. Origins of Antibodies — *Ted Steele*

Saturday 16th April 2016

Roebrooks Farm (Melaleuka Stud), Nth Dandalup
AI and Embryo Transfer Demonstration, Dom Bayard

Retreat Participants:

Professor Roger L Dawkins, Chairman, CYO Foundation
Mr Bruce Cheung, Pardoo Cattle Station, Kimberly, WA
Mrs Barbara Roberts-Thomson, Tevanion, nr Armidale, NSW
Mr Dom Bayard, Global Reproductive Solutions, Goorambat, Vic
Mr. Peter Gilmour, Irongate Wagyu, Albany, WA
Bob and Jan Reed, Mazda Wagyu International, Mossiface, Vic
Ms Sharon Baker, Executive Administrator, TMG, CYO Foundation
Mr Anthony Lloyd, Vice Chairman, CYO Foundation
Ms Sue Lester, Rheumatology Dept., Queen Elizabeth Hospital, SA
Dr. Sally S Lloyd, Researcher, CYO Foundation
Dr. Edward J Steele, Honorary Research Associate, CYO Foundation
Dr. Jose Valenzuela, CYO Foundation and Melaleuka Stud
Ms Kelly Manning, Melaleuka Stud
Dr. M Luisa Ashdown, Patents & Licensing, Vic
Kelly Manning, Melaleuka Stud, Nth Dandalup
Mr John R Dawkins, Tilbrook Melaleuka Group
Mr Alan Peggs, Pastoral Consultant, WA/ Pardoo Station
Assoc Professor Jerzy K Kulski, University of Western Australia
Mr Roger Stubbs, CYO Foundation
Mr Ed Turner, CYO Foundation Board
Mr Patrick Dawkins, CYO Foundation Board
Mr Jason Wallis, CYO Foundation Board
Mr Ben Dawkins, CYO Foundation Board
Mr Brian Walker, CYO Foundation Board
Mr Ian Oldfield, CYO Foundation Board
Mr Wayne Cox, CYO Foundation Board
Mr Renny Lee Steere, CYO Foundation Board
Dr. Christine Storer, Adjunct Professor, Charles Sturt University, Agribusiness
Ms Wendy Dymond, Lecturer Charles Sturt University, Agribusiness
Professor Peter Sharp, Professor & Director, Plant Breeding Institute, University of Sydney
Mr. Ray Candy, Hunter Irwin & Co, Seed Merchants

Mr. Rob Giura, Mucca Dairy Farming, Nth Dandalup
Mr. Adrian Pitsikas, Greenacres Turf, Serpentine
Professor Mike Jones, Murdoch University
Mr. David Arkwright, Peel Regional Commission
Mr. Mr. Dean Unsworth, Murray shire, CEO
Mr. Brett Flugge, Murray shire Executive Manager
Mr. Jonathan Emanuel, Emanuel Brothers, Beef Livestock
Mr. Peter Bates, Peel Beef Farming
Mr. Bruce Campbell, Peel Livestock Farming
Mr. Geoff Manning, Peel Dairy Farming
Mr Ted Edwards, Agriculture BSc Hons student
Mr Lindsay Baker, Manager, Melaleuka Stud
Mr Martin Ashdown, Biotempus Pty Ltd, Melbourne
Mr Tom Edwards, BSc Hons Student, Murdoch University

28th-29th July: Future Proofing

Rose Cottage, Melaleuka Stud, North Dandalup

Symposium 1: Healthy Beef

Pure or full, genetic improvement, Aka's in or out *Roger Dawkins*
Tm and Haplotypes in Akaushi, South Devon, Senepol *Sally Lloyd*
New Projects – testing for deficiencies *Jose Valenzuela*
Synteny
Louise Smyth/Sally Lloyd

Symposium 2: Healthy Beef

History and place of reds in Australia and the Americas Bob Reed
Meeting the future *Keith Hammond*
Embryo selection *Dom Bayard*
IVF and Embryo Biopsy *Brendan Choo*
Update Pardoo: Red versus Black Wagyu, preferred F1 *Alan Peggs*
Preferred crossbreds *Darrin Hamblin*

Symposium 3: The CYO Pasture & Kikuyu Research Program

Results – Kikuyu research program *Tony Lloyd, Adrian Pitsikas*

Symposium 4: Diseases

BJD and Crohn's *Roger Dawkins/Dom Bayard*
Enhancing immunity in dogs and cattle *Martin Ashdown*
The Power of IP *Louisa Ashdown*

Symposium 5: The Needs in Peel

Local Abattoir *Rob Giura*
Abattoir Waste Management *Gary Watson*
Genetics for export *Dom Bayard*

Round table: The Heartbrand World
led by Roger Dawkins and Bob Reed

Conference Dinner Sebel, Mandurah
Key Speakers:

Community Involvement	*Tom Dawkins*
Transforming Peel	*Norm Baker*

Friday 29th July 2016

Tour of Melaleuka Stud and Judging of Red Wagyu

Breeders Clinic	Bob Reed/Keith Hammond/Dom Bayard

Symposium 6: Feedlots

Feeding Cattle	Dean Maughan
Challenges for Feedlots	Roger Dawkins

Retreat Participants:

Professor Roger L Dawkins	CYO Foundation (and Melaleuka Stud)
Mr Dom Bayard	Global Reproductive Solutions, Goorambat, Vic
Bob and Jan Reed	Mazda Wagyu International, Mossiface, Vic
Mrs Sharon Baker	CYO Foundation
Mr Anthony Lloyd	CYO Foundation
Dr. Sally Lloyd	CYO Foundation
Dr. Jose Valenzuela	CYO Foundation and Melaleuka Stud
Dr. Luisa Ashdown	CYO Foundation Board
Mr John R Dawkins	Tillbrook Melaleuka Group
Mr Alan Peggs	Pastoral Consultant, WA/ Pardoo Station
Mr Patrick Dawkins	CYO Foundation Board
Mr Ben Dawkins	CYO Foundation Board
Mr Wayne Cox	CYO Foundation Board
Mr Renny Lee Steere	CYO Foundation Board
Mr. Rob Giura	Mucca Dairy, Nth Dandalup-CYO Foundation Board
Mr. Adrian Pitsikas	Greenacres Turf, Serpentine
Mr. Bruce Campbell	Peel Livestock Farming
Mr Lindsay Baker	Melaleuka Stud
Mr Martin Ashdown,	Biotempus, Melbourne
Dr. Louise Smyth	University of Notre Dam

6th October – Candidate presentations

Genomic Basis of non-progressive HIV disease	Nitin Saksena
Molecular characterization of the herbicide resistance	Wendy Liu
Genetic modification of flower colour	Tim Holton
Selecting for Low Tm Beef	Sally Lloyd

12th -15th December CYO Laboratory Course and Retreat

12th December Kielman Laboratory

9:00-12:00

Presentation: Introductions, Course Overview, Laboratory Orientation and Safety Sally Lloyd / Jose Valenzuela

Practical: Simultaneous extraction of Tm and DNA from meat part 1 Sally Lloyd / Jose Valenzuela

Practical: Measurement of DNA concentration and preparation of dilution Sally Lloyd / Jose Valenzuela

Presentation: IP and research licences RL Dawkins

Presentation: Steps to a patent Glen McIntyre

13:00-17:00

Practical: Polymerase Chain Reaction Sally Lloyd / Jose Valenzuela

Practical: Fat and DNA extraction part 2 Sally Lloyd / Jose Valenzuela

Practical: Tm measurement part 1 Sally Lloyd / Jose Valenzuela

13th December Kielman Laboratory

9:00-12:00

Practical: Alcohol Precipitation of DNA

Practical: Tm measurement part 2

Practical: Measurement of DNA concentration and quality

Practical: Capillary Electrophoresis on Qiaxcel

Practical: Analysis of Tm measurement

13:00-17:00

Practical: Analysis of Electrophoresis results

Workshop: Using PCR Markers to determine haplotype segregation in a pedigree

18:00 Rose Cottage BBQ and Presentation of Certificates

Course Participants

Samantha Connolly, Sarah Hamblin, Sarah Hammond, Paula Pownall, Glen McIntyre

14th December, Kielman Laboratory Seminar Room

Symposium 1: Incubating industries in the Peel Region

Panel - Introductions

CYO'Connor Foundation Business Incubators Roger Dawkins/ Glen McIntyre
"Transform Peel" How it can Support Innovation Linda Cunningham
Melaleuka Stud: Experiments in Agriculture Lindsay Baker
Endpoints for Pardoo Research support Alan Peggs

Project Applicants: engineering

Modular Solar Irrigation Rob Giura
Removing iron with PEFT Hilary Nath
The role of electric bikes in sustainable urban developmentJulian Ilich

Project Applicants: Cattle

Genetic selection for polling in cattle Roger Dawkins
Choosing pastures for marbled beef Brad Nutt
Nutrient stripping using Kikuyu Tony Lloyd
Provenance testing for meat from paddock to plate John Dawkins
Accurate Tm and fat monitoring in live cattle Jose Valenzuela

Services supporting innovative business in Peel

Internet Delivery to Peel Region Terry Filipo
Engineering and fabrication solutions Daniel Jones
Feed Nutrition testing John Milton

Project Applicants:

Novel techniques for cold pasteurisation of milk Hilary Nath
Point of sale Tm and DNA testing Sally Lloyd
Grubs up: Edible insects, feasibility study Paula Pownall

Project Applicants: pasture

Need for genetic testing in Rye Grass Ray Candy
BACD in Rye grass: Results Pilot Study Sally Lloyd
Soil improvement in perennial pasture. Andy Gulliver (CWise)
Improved pasture productivity with silage Geoff Manning
Optimising carbon sequestration in pasture Richard Harper
Overplanting to improve winter production of Kikuyu pasture Ray Candy

15th December Kathleen Room CYO'Connor Village Pub

Symposium: Research into Healthy Beef and Pasture

Role of adrenal and insulin regulation in fat deposition Peter McGilchrist
Using High Fibre technology to efficiently produce high quality beef Dean Maugham
Feed quality of pasture John Milton
RNAi in the field Glen McIntyre
DNA editing technologies for agriculture Michael Jones
Feeding strategies for Wagyu T Suzuki (remote)
Genetic influences controlling fat softening in response to feeding Sally Lloyd
Biopsy to monitor fat deposition Jose Valenzuela
Veterinary Training to Support Agriculture Peter Irwin
Evaluating animals and carcasses for meat quality Phil Petrovich

Symposium: Synteny and Synergies: cross fertilisation of human health and animal production research

Syntenic regions regulate Muscular Dystrophy and Beef Marbling Roger Dawkins
A Common Platform to treat inherited diseases Steve Wilton
Histochemistry of Marbling, Dystrophy and Statin induced Myositis Louise Smyth

Workshop: Patenting the Bovine Genome

Panel: RL Dawkins, SS Lloyd, AJ Lloyd, J Valenzuela, S Wilton
Branhaven and Cargill Patent RL Dawkins/SS Lloyd
CY O'Connor Foundation patents RL Dawkins/SS Lloyd/J Valenzuela
Response and Commentary
S Wilton, G McIntyre, P McGilchrist, M Jones Dave Morrison

17:00 – 19:00 Dawkins Family Christmas Party – The Vines, CY O'Connor Village Pub

17:30** CYO Foundation Board Meeting – Kathleen Room, **CY O'Connor Village Pub

Abstracts: December Retreat

Offering Engineering Design and Fabrication Solutions

Daniel Jones, Pure Engineering

Pure, formed in 2006 is unique, it specialises in custom machinery for a variety of industries including:

i) Electric hospital bariatric bed movers;
ii) Door stacking machines;
iii) Pharmaceutical wall passes through conveyor with air lock,
iv) Intermediate Bulk Container washing machines.

The business has been focused on automation in industrial applications it endeavours to broaden its horizons, relishing the challenge of achieving excellence in new areas.

Pure is grateful for the opportunity to offer its services to Peel industries.

It was suggested Pure could offer a 'calf catcher', currently only available for import. Pure is confident it can offer a design with all the available features and more convenient attachment and operation.

Veterinary Training in Animal Production

Peter Irwin, Murdoch University, College of Veterinary Medicine

Students enrolled in the veterinary science course at Murdoch University undertake five years of study. The first year of the course is closely aligned with animal science and students commence their farm placements at the end of this year. Animal production systems and farm safety are key aspects of their early training. Veterinary students learn about the health and diseases of all species during the course and receive intensive instruction about the management and treatment of the major farm animal species (cattle, pigs, small ruminants and poultry). Clinical placements once more bring them onto farms, feedlots, small holdings and extensive properties as they complete their clinical training. Murdoch University graduates approximately 85 new veterinarians each year.

A Common Platform to treat inherited diseases

Steve Wilton, Centre for Comparative Genomics, Murdoch University.

Duchenne muscular dystrophy (DMD), a common and severe form of childhood muscle wasting, is caused by mutations in the dystrophin gene. We have developed a novel therapy by redirecting the dystrophin pre-mRNA splicing and excise specific exons from the mature mRNA. "Exon skipping" has progressed from a concept to Phase 3 clinical trials in less than two decades. One compound: Exondys51 has been granted accelerated approval by the US FDA, showing that therapeutic alternative splicing can alter gene expression at clinically relevant levels. We propose that splice switching oligomers could be developed as personalized genetic therapies to treat many disease-causing mutations.

Non-thermal electro-pasteurisation by low DC voltages using PEFT cell

Hilary Nath, EcoAdvise

Non-thermal electro-pasteurisation has been achieved by low DC voltages using a perforated electrode flow through (PEFT) cell with micro-gap electrode spacing. The titanium anode of the cell was partially insulated with a titanium oxide nano film allowing an electrode gap to be reduced to 40μm. Anodic insulation also reduced current flow and ohmic heating. Microbial inactivation was investigated using a gram-negative Escherichia coli bacteria suspended in a chloride-free electrolyte having a conductivity of 4.2 mS/cm mimicking a fluid food sample. Inactivation commenced at field strengths of approximately 10kV/cm, generated by 40 V DC applied across the electrodes. A six log inactivation was achieved by 20 hydrodynamically controlled pulses of an applied field of 22.5 kV/cm generated by 90 V DC consuming a power of 430 J/mL, which under continuous operation without cooling, gave a maximum temperature at 42°C.

Key Words: Micro-gap, perforated electrode, non-thermal electro-pasteurisation, titanium oxide nano film.

Iron removal from groundwater using PEFT cell

Dr. Hilary Nath, EcoAdvise

An electrochemical method has been developed to provide an effective means of achieving rapid oxidation of iron in contaminated groundwater. It is based upon a novel perforated electrode flow through (PEFT) cell using titanium electrodes coated with a RuO_2/IrO_2 catalytic film. Electro-oxidation of natural groundwater with a conductivity of 350 μS/cm containing 6 mg/L of iron achieved 99.8% iron removal (residual total iron 0.01 mg/L) at the power consumption of 0.13 kWh/m^3 operating at a flow rate of 3 m^3/day.

Key Words: Electro-oxidation, perforated electrode, iron removal, groundwater

Re-engineering sandy soils to increase soil carbon storage: a new climate mitigation technology

R. J. Harper and S.J. Sochacki, School of Veterinary and Life Sciences, Murdoch University

Increasing the storage of soil organic carbon (SOC) is a means of mitigating increasing atmospheric concentrations of carbon dioxide. Net changes in SOC storage are dependent on maintaining a new land management system but this can limit flexibility in land use and also require continuous inputs. In natural soils SOC content is often positively correlated with clay content thus additions of clay to sandy soils may increase soil carbon storage. This proposition was tested and found to hold in six long-term (19-29 year old) experiments on the Swan Coastal Plain. Bauxite processing residue had been applied 19-29 years previously at rates of 40 to 2000 t/ha to annual pastures and irrigated horticultural sites with sandy Arenosols. Changes in carbon sequestration in the surface 0.3 m ranged between 10 and 82 t CO_2-e/ha, or 2-4 t CO_2-e/ha/yr across all sites. Clay additions also resulted in co-benefits such as increased agricultural yields, and reductions in wind erosion, water repellency and leaching of nutrients and pesticides. Not only does this represent a new climate mitigation technology but it will allow the agricultural intensification of the Swan Coastal Plain sandy soils which occur in abundance.

RNAi in the Field *Dr. Glen McIntyre*

RNA interference (RNAi) is a biological process in which dsRNA molecules inhibit gene expression, typically by causing the destruction of specific mRNA molecules. Expressing double-stranded RNA (dsRNA) in transgenic plants to silence essential genes within herbivorous pests is referred to as trans-kingdom RNA interference (TK-RNAi). TK-RNAi has emerged as a promising strategy for crop protection, with the potential to augment or supplement traditional pesticide methods. However, the natural processing of dsRNA into siRNAs by the plant's intrinsic RNAi machinery may reduce this pesticidal activity. Detailed here is a snapshot of work conducted to subvert the natural dsRNA processing within the host plant by exploiting sub-cellular compartmentalisation in chloroplasts. This work has enabled the stable accumulation of introduced dsRNA in plants in a way that provides effective protection against herbivorous pests without consequence to normal plant functions. This work is now the subject of international patents and holds the promise of providing new, environmentally friendly, and durable pest resistance in crops.

Made in the USA
Monee, IL
05 November 2026

67379774R00040